SpringerBriefs in Applied Sciences and Technology

SpringerBriefs present concise summaries of cutting-edge research and practical applications across a wide spectrum of fields. Featuring compact volumes of 50–125 pages, the series covers a range of content from professional to academic.

Typical publications can be:

- A timely report of state-of-the art methods
- An introduction to or a manual for the application of mathematical or computer techniques
- A bridge between new research results, as published in journal articles
- A snapshot of a hot or emerging topic
- An in-depth case study
- A presentation of core concepts that students must understand in order to make independent contributions

SpringerBriefs are characterized by fast, global electronic dissemination, standard publishing contracts, standardized manuscript preparation and formatting guidelines, and expedited production schedules.

On the one hand, **SpringerBriefs in Applied Sciences and Technology** are devoted to the publication of fundamentals and applications within the different classical engineering disciplines as well as in interdisciplinary fields that recently emerged between these areas. On the other hand, as the boundary separating fundamental research and applied technology is more and more dissolving, this series is particularly open to trans-disciplinary topics between fundamental science and engineering.

Indexed by EI-Compendex, SCOPUS and Springerlink.

More information about this series at http://www.springer.com/series/8884

Rabiu Muazu Musa · Zahari Taha
Anwar P. P. Abdul Majeed
Mohamad Razali Abdullah

Machine Learning in Sports

Identifying Potential Archers

Springer

Rabiu Muazu Musa
Innovative Manufacturing, Mechatronics and Sports Laboratory, Faculty of Manufacturing Engineering
Universiti Malaysia Pahang
Pekan, Pahang, Malaysia

Anwar P. P. Abdul Majeed
Innovative Manufacturing, Mechatronics and Sports Laboratory, Faculty of Manufacturing Engineering
Universiti Malaysia Pahang
Pekan, Pahang, Malaysia

Zahari Taha
Innovative Manufacturing, Mechatronics and Sports Laboratory, Faculty of Manufacturing Engineering
Universiti Malaysia Pahang
Pekan, Pahang, Malaysia

Mohamad Razali Abdullah
Universiti Sultan Zainal Abidin
Kuala Terengganu, Terengganu, Malaysia

ISSN 2191-530X ISSN 2191-5318 (electronic)
SpringerBriefs in Applied Sciences and Technology
ISBN 978-981-13-2591-5 ISBN 978-981-13-2592-2 (eBook)
https://doi.org/10.1007/978-981-13-2592-2

Library of Congress Control Number: 2018955442

This Springer imprint is published by the registered company Springer Nature Singapore Pte Ltd.
The registered company address is: 152 Beach Road, #21-01/04 Gateway East, Singapore 189721, Singapore

This book is dedicated to my beloved family late Malam Muazu Musa, Halima Muazu, Adamu Muazu, Sadiya Muazu and Abubakar Muazu as well as my precious wife Nafisa Usman Yahaya.

Rabiu Muazu Musa

I would like to dedicate this book to my family, P. P. Abdul Majeed K. Idros, Sulleha Eramu, Noriza P. P. Abdul Majeed, as well as my wife Sharifah Maszura Syed Mohsin and not forgetting the apple of my eyes, Saffiya Anwar.

Anwar P. P. Abdul Majeed

Acknowledgements

We would like to acknowledge Professor Dr. Zahari Taha and Associate Professor Dr. Mohamad Razali Abdullah for providing us with the necessary knowledge and valuable suggestions for making the realisation of the book possible. We would also like to extend our gratitude to UniSZA sports scientists, especially Ahmad Bisyri Husin Musawi Maliki, Siti Musliha Mat-Rashid, Norlaila Azura Kosni, Norzulaika Alias and Vijayamurugan Eswaramoorthi for their support in carrying out the data collection for this project. Our heartfelt appreciation is extended to the Innovative Manufacturing, Mechatronics and Sports Laboratory (iMAMS) members (Azraai, Hasnun, Yashim, Fakri, Aizzat, Jessnor, Amirul, Qastalani, Fahriza, Ash, Yusri, Ismail and Shahrizan) for making our stay at the laboratory a memorable one. We would also wish to acknowledge the Terengganu Sports Council (MSN) and the entire coaching staffs of the Terengganu archery programme for their support in the accomplishment of this project. Moreover, we would also like to extend our sincere appreciation and gratitude to the National Sports Institute of Malaysia (ISN) in awarding the grant for this study.

Rabiu Muazu Musa
Anwar P. P. Abdul Majeed

Contents

About the Authors

Dr. Rabiu Muazu Musa holds a Ph.D. degree from Universiti Sultan Zainal Abidin (UniSZA), Malaysia. He obtained his M.Sc. in sports science from UniSZA in 2015 and his B.Sc. in physical and health education at Bayero University Kano, Nigeria, in 2011. His Ph.D. research focused on the development of multivariate and machine learning models for athletic performance. His research interests include performance analysis, health promotion, sports psychology, exercise science, talent identification, test and measurement as well as machine learning.

Prof. Dr. Zahari Taha graduated with a B.Sc. in aeronautical engineering with honours from the University of Bath, UK. He obtained his Ph.D. in dynamics and control of robots from the University of Wales Institute of Science and Technology in 1987. He is currently a Professor of Manufacturing Engineering at Universiti Malaysia Pahang, and he is also the founder and advisor of the Innovative Manufacturing, Mechatronics and Sports (iMAMS) Laboratory, Universiti Malaysia Pahang. He teaches and conducts research in the areas of industrial automation, robotics, ergonomics, sustainable manufacturing, machine learning and sports engineering.

Dr. Anwar P. P. Abdul Majeed graduated with a first-class honours B.Eng. in mechanical engineering from Universiti Teknologi MARA (UiTM), Malaysia. He obtained an M.Sc. in nuclear engineering from Imperial College London, UK. He then received his Ph.D. in rehabilitation robotics under the supervision of Prof. Dr. Zahari Taha from Universiti Malaysia Pahang (UMP). He is currently serving as a senior lecturer at the Faculty of Manufacturing Engineering, UMP. He is an active research member at the Innovative Manufacturing, Mechatronics and Sports Laboratory, UMP. His research interest includes computational mechanics, applied mechanics, sports engineering, renewable and nuclear energy, sports performance analysis as well as machine learning.

Assoc. Prof. Dr. Mohamad Razali Abdullah obtained his Bachelor of Physical Education in 1989 from Universiti Putra Malaysia (UPM). He obtained his M.Sc. in sport and exercise science from the University of Wales Institute, Cardiff, in 1998, and in 2007, he received his Ph.D. in sports science from UPM. His research interests include motor control, sports biomechanics, motor performance and machine learning in sports. He is currently an Associate Professor at Faculty of Applied Social Sciences, Universiti Sultan Zainal Abidin, Kuala Terengganu, Malaysia.

Chapter 1
Introduction

1.1 Nature of Archery

The sport of archery constitutes typically the act of shooting an arrow with a bow to a predetermined target. The sport could also be described as a static and a close skill that involves a face-to-face shooting competition within a specified period of time. A standard archery competition comprises of a number of events including an individual archer qualifying event, head-to-head match event as well as three-person team match play [1]. It is important to note that the separations of the contestants in a competition are customarily carried out into one of the three aforesaid categories with regard to gender and the type of bow, i.e. recurve or compound bow.

1.2 Performance Parameters Influencing Archery Performance

Successful performance in the sport of archery is reliant upon a myriad of factors. In essence, performance in this sport depends on the interactions of several performance indicators. Thus, for an archer to excel in the sport, a proper preparation with regard to all the encompassing variables should be performed prior to and during a competition. Performance variables such as physical fitness consisting of stamina and motor ability, physiological indicators involving nutrition, heart rate, blood pressure and psychological elements comprising of concentration, mental skills, confidence and motivation as well the anthropometrical attributes of the athletes have been previously reported as the major determinants that could influence performance in the sport [2–5].

In many archery programmes and organisations, managers as well as coaches strive to recognise the overall performance elements that could influence performance in the sport. For this reason, sports scientist and sports professionals attempt

R. Muazu Musa et al., *Machine Learning in Sports*, SpringerBriefs in Applied Sciences and Technology, https://doi.org/10.1007/978-981-13-2592-2_1

to provide information that could be used to guide decision-making and optimise performance. Although several attempts have been made in view to offering solutions by means of studying the interaction of several performance variables for the delivery of performance in the sport, such solutions are more or less subjected to conventional analysis that often fall short in addressing the problem. It is against this background that the present brief is initiated to provide an avenue for performance evaluation through the employment of a different approach that could be beneficial to coaches, sports managers and other stakeholders in the domain. The brief is also directed towards the establishment of a unique technique for the identification of potential archers which has been commonly a rigorous task for the people involved in the talent identification programmes.

1.3 Machine Learning in Sporting Activities

The employment of machine learning and its derivatives in the domain of sporting activity and human movement is not entirely unusual mainly owing to the advancement of disruptive technologies. This subsection shall briefly review on the attempts made by researchers in applying machine learning in the aforesaid domains. It is worth to note that the list is not exhaustive, but it provides a general overview of its vast usage in prediction as well as classification. Artificial Neural Networks (ANNs) have been utilised in the prediction of energy expenditure (EE) based on raw accelerometer data attained from different parts of the body, i.e. wrist, thigh and hip [6]. It was shown from the study that the ANN model provided a better prediction of the EE in comparison to linear regression as well as linear mixed models.

Ellis et al. [7] used a random forest classifier to predict EE as well as to classify a number of physical activities (PA) by through heart rate (HR) and accelerometer data. In addition, the authors utilised random forest regression trees to predict metabolic equivalents (MET). Different time- and frequency-domain features were extracted and normalised to provide an unbiased distribution of the features that may influence the classification accuracy as well as the predictability of the machine learning models. The leave-one-out cross-validation technique was adopted in the study. It was shown that both the regression and classification models evaluated were able to yield desirable prediction of the MET as well as a reasonable classification of the PA. It was also demonstrated that the inclusion of the HR data improved the MET estimation; however, the inclusion does not improve the classification of the activity types significantly.

A comparative assessment of the classification of nineteen different human daily and sporting activities with different machine learning algorithms was evaluated by Barshan and Yüksek [8]. Inertial and magnetic sensors, i.e. accelerometer, magnetometer as well as gyroscope, were placed on the chest, arms and legs to extract the features. Principal component analysis (PCA) was used to identify the dominant features for the investigation. The study is rather extensive as it evaluated different cross-validation techniques apart from investigating the effect of combining different sensors towards the classification efficacy of the models evaluated. It was shown that ANN, Support Vector Machine (SVM) and Gaussian Mixture Model (GMM) provided classification accuracy rates of 99.2, 99.2 and 99.1%, respectively.

Acceleration signals attained from the wrist and hip were used to classify different activities, viz. lying down, standing, walking, sitting, running, dancing as well as basketball via logistic regression (LR) model [9]. A threefold cross-validation technique was used for the study. It was shown that the hip-based model classifies better in comparison to the wrist-based model with respect to standing, basketball, running and lying down. Conversely, the wrist model classifies better with regard to sitting and dancing, whilst identical classification performance is exhibited for walking.

Machine learning has also been used to predict match outcomes in different sports. Joseph, Fenton and Neil evaluated the prediction capability of different machine learning models, namely expert Bayesian Network (eBN) against decision tree learner (MC4), Naive Bayes (NB), data-driven Bayesian and k-Nearest Neighbour in determining the match outcome for Tottenham Hotspur Football Club [10]. The features selected for the study are the availability of the key players evaluated at that time, the playing position of a particular player playing in the midfield, the quality of the opposing team, the quality if the Spurs attack, the quality and performance of the Spurs' team as well as the venue, viz. home or away. The test data and training data were separated across the seasons. It was shown that the eBN provided a more accurate representation of the win, lose or draw for Spurs with an accuracy of 59.21% against other evaluated models.

The employment of ANN and LR has also been investigated in predicting English Premier League (EPL) results [11]. A number of features were selected, namely home and away goals, Home and away shots, home and away corner, home and away odds, home and away attack strength, home and away players' performance index, home and away managers' performance index, home and away managers' win, and home and away streak. The data was extracted from a total of 110 matches played in the 2014/2015 EPL season, in which 90 sets were used for training and 20 sets of matches were used to evaluate the predictive ability of the models evaluated. It was demonstrated that ANN and LR models are able to yield a classification accuracy of 85% and 93%, respectively. However, it is worth to note that the classification class evaluation is different for both models, i.e. ANN model classifies win, lose or draw, whilst the LR model classifies winning or losing, hence a direct comparison could not be made.

A Gaussian-based SVM has also been investigated in predicting match results of EPL data in an extended study [12]. A similar methodology was undertaken with respect to the feature selection as in [11]. It was reported that the classification accuracy attained through the employment of the algorithm is 53.3% in classifying win, lose or draw suggesting that the proposed algorithm is not suitable for the data.

Razali et al. [13] applied BN to predict EPL match results by taking into account three consecutive seasons from 2010 to 2013. The features selected are home team goals, away team goals, home team goals shots, away team goals shots, home team shots on target, away team shots on target, home team corner, away team corner, home team foul committed, away team foul committed, home team yellow card, away team yellow card, home team red card and away team red card. A tenfold cross-validation technique was utilised on the extracted data set, and it was reported that the classification accuracy attained by means of the proposed algorithm is 75.09%.

ANN has also been used to predict team league's rankings in volleyball in Turkey [13]. The data was taken from 66 matches between the year 2013 and 2015. In this regression study, a network topology of 4-4-1 is employed by utilising the logsig activation function. The input parameters selected were wins, defeats, home wins and away wins, whilst the output parameter is the team ranking. The data was normalised, and the ratio for training, testing and validation selected for the investigation is 70:15:15. It was shown that the ANN model is able to provide a coefficient of determination with a value of 0.985 between the actual and predicted data suggesting the efficacy of the ANN model in predicting team rankings.

The employment of ANN has also been applied to identify the important elements (features) that ascertain the result of basketball games [14]. The data was taken from Serbia's First B basketball league for men for five seasons with 890 game results were analysed. It was shown from the study that with the selection of appropriate features, namely two-point shots under the hoop and defensive rebound, the classification accuracy of the developed ANN model in classifying the winning and losing teams might be increased from 66.4 to 80.96%.

The utilisation of machine learning has also been investigated on non-team sports. Nonlinear regression models were compared against ANN in predicting the distance of javelin throws [15]. A number of predictors were ascertained through statistical means, namely cross step, specific power of the arms and the trunk, specific power of the abdominal muscles and grip power. A network topology of 4-3-1 was used in the study upon judicious testing. The data set used for learning, validation and testing were 40, 15 and 15 cases, respectively. Additional fresh data of 20 cases was used for further verification on the prediction efficacy of the developed models. It was demonstrated that the ANN was able to provide a better prediction on the throw in comparison of the regression model.

Edelmann-Nusser et al. employed ANN in predicting the performance of an elite female swimmer in the finals of the 200-m backstroke at the 2000 Olympic Games in Sydney [16]. A total number of 19 data sets were utilised in the study, and owing to the limited data available, the leave-one-out cross-validation technique was used. The 10-4-1 network topology was used, with the input being a number of selected features whilst the output is the swim time. It was shown that the model provided an accurate prediction with an error of 0.05 s in comparison to the linear regression analysis that provided unsatisfactory results.

It is evident from the literature presented that the utilisation of machine learning has been extensively reported in the domain of physical activities as well as sports in general. However, it is worth noting that the employment of such technique in determining the performance of a given athlete in a given sport with respect to different performance features is limited. This brief therefore attempts on addressing this gap, particularly with regard to the sport of archery.

1.4 *K*-Means Clustering

In this brief, the k-means clustering algorithm, which is essentially a type of unsupervised learning, is used to separate the classes of the related performance variables assessed. It has been reported in the literature that the k-means clustering method is more reliable than the hierarchical agglomerative clustering [17]. This is primarily due to the fact that it operates on actual observations rather that dissimilarity measures employed in hierarchical clustering. The number of clusters, k is selected to be two as the data consists of HPA and LPA. The k centroids are then randomly initialised, and the instances that are closest to the centroid are assigned through the minimisation of the sum of distances from all the data in the cluster.

1.5 Classification Algorithms

1.5.1 K-*Nearest Neighbour (*K-*NN)*

The k-Nearest Neighbour (k-NN) is a nonparametric regression and classification method established by Fix and Hodges in the fifties [18]. It is often considered as one of the simplest types of supervised machine learning algorithms. However, k-NN did not attain appreciable attention until the sixties primarily due to the inadequate computing power prior to it. This simple machine learning scheme is also recognised as 'lazy learning' or 'instance-based learning' as it does not require learning, in other words, the computation of the algorithm occurs during runtime [19]. k-NN has been successfully applied for classification in a different number of diverse disciplines [20–23].

The number of neighbours, k for fine and coarse are 1, and 100 whilst the remaining variation i.e. medium, coarse, cosine, cubic and weighted, the number of neighbours is selected to be 10. As for the distance metrics, Euclidean distance is utilised in the fine, medium, coarse and weighted k-NN variations, whilst the cosine and cubic employ the cosine and a special case of the Minkowski distance, respectively. The 'no distance weight' (i.e. equal distance) was used for all variations except the weighted k-NN, in which the weight is the squared inverse of the distance. The distance formulae for Euclidean (d_e), cosine (d_c) and cubic (d_3) are as follows:

$$d_e(x_i, x_j) = \sqrt{\sum_{r=1}^{n} \left|(x_i) - (x_j)\right|^2}$$

$$d_c(x_i, x_j) = 1 - \frac{x_i . x_j}{\sqrt{(x_i . x_i).(x_j . x_j)}}$$

$$d_3(x_i, x_j) = \sqrt[3]{\sum_{r=1}^{n} |(x_i) - (x_j)|^3}$$

where d is the distance between new data point x_i and training data point x_j.

1.5.2 Support Vector Machine (SVM)

SVM is utilised to acquire the optimal hyperplane that correctly classifies the multi-class classification problem, i.e. the performance of the archers, namely LPA and HPA. The acquisition of the optimal hyperplane is via the identification of the maximal distant from the classes that in turn minimises the risk of misclassification of both the training and validation data set.

The generalised SVM classification function is given by

$$f(x) = \sum_{i=1}^{l} y_i \alpha_i K(x_i, x) + b$$

whereby $K(x_i, x)$ is the kernel function that is used to measure the training vector (x_i, x). Different kernel functions are used in this study to investigate its efficacy in classifying the data correctly. The kernel functions employed are

$$K(x_i, x) = \left(x_i^T.x\right), \text{ Linear kernel}$$

$$K(x_i, x) = \left(\frac{X_i^T \cdot X + 1}{\mathrm{C}}\right)^d, \text{ Polynomial kernel of degree } d$$

$$K(x_i, x) = \exp\left\{-\frac{(X - X_i)^2}{\sigma^2}\right\}, \text{ Radial Basis Function (RBF) kernel}$$

The following objective function is minimised in order to obtain the optimal classification function as follows:

$$\min \frac{1}{2} w^T w + \mathrm{C}\left(\sum_{i=1}^{l} \varepsilon_i\right)$$

where C is the penalty parameter, whilst ε is the slack variable. A more in-depth mathematical treatment of the subject matter could be found in [24]. In this study, six different kernel functions are investigated, namely linear (SVM-L), quadratic (SVM-Q), cubic (SVM-C), fine RBF (SVM-F-RBF), medium RBF (SVM-M-RBF) and Coarse RBF (SVM-C-RBF). The scale of the fine, medium and coarse RBF is defined by $0.25^* P^{\frac{1}{2}}$, $P^{\frac{1}{2}}$ and $4^* P^{\frac{1}{2}}$, respectively, where P is the number of predictors. The solver used for the training is the sequential minimal optimisation algorithm.

1.5.3 Artificial Neural Network (ANN)

ANN is a popular machine learning algorithm primarily due to its ability to cater for nonlinear behaviour of the data analysed. The model is inspired by the biological nature of the nervous system [25]. The model's output is reliant on the activation function that stipulates the threshold value and its input as well as the weights that are associated with the input (x). It is worth noting that the weight (w) defines the strength of the input whilst the bias (b) controls input for the activation (transfer) function. Different types of activation functions have been investigated, a more elaborate discussion of the types of activation functions may be found in [25, 26]. A general expression of the output (u) model may be expressed as follows, in which the output model is then governed by the activation function selected.

$$u = b + \sum_{i=1}^{N} w_i x_i$$

Typically, an ANN model consists of an input layer, a hidden layer as well as the output layer. The number of neurons of the input layer is determined based on the desired input and output parameters, for instance in this brief, the input is governed by a myriad of performance indicators whilst the output is either HPA or LPA. The number of neurons in the hidden layer is often determined heuristically.

1.5.4 Logistic Regression (LR)

Logistic regression is a statistical modelling method that is based on the notion that a sigmoidal relationship exists between the probability of group membership as well as the predictor variables [27–30]. The logistic model may be represented via the following equations:

$$z = \beta_0 + \sum_{i=1}^{n} \beta_i x_i$$

$$P(z) = \frac{e^z}{1 + e^z}$$

where z is a measure of the influence of the descriptive variables x_i ($i = 1, \ldots, n$), β_i are the regression coefficients which are obtained by maximum likelihood in conjunction

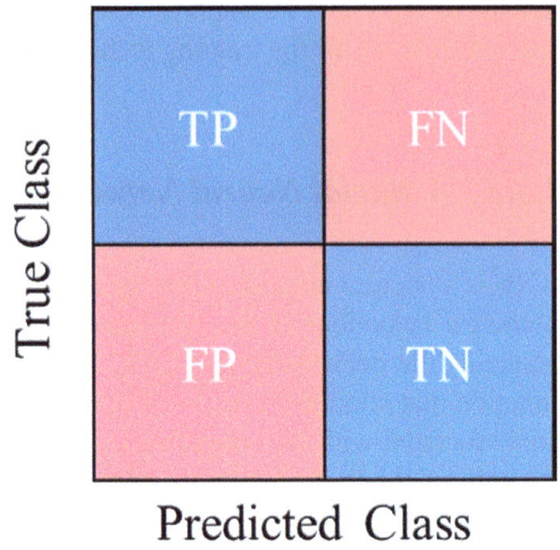

Fig. 1.1 Confusion matrix

with their standard errors $\Delta\beta i$, and P(z) is the categorical response of the variables that represent the probability of an archer to be categorised as HPA or LPA.

1.5.5 Model Performance Evaluation

The classification models evaluated in this brief are assessed by means of classification accuracy (ACC), specificity (SPEC), precision (PREC), sensitivity (SENS), error rate (ERR) as well as Matthew's correlation coefficient (MCC). The ACC is essentially the ratio between the number of correctly classified observations and the total number of observations. The SENS and the SPEC are the true positive rate or the positive class accuracy as well as the true negative rate or negative class accuracy, respectively. The PREC computes the number of correct positive predictions over the total number of positive predictions. The ERR, on the other hand, appraises all misclassifications over the number of total observations. Conversely, the MCC measures the quality of binary classification, and it has a range from -1 to 1 whereby 1 suggests an entirely accurate binary classifier. The confusion matrix (Fig. 1.1) allows the observation of correctly classified and misclassified observations that transpires between the defined classes. The detailed method of acquiring the aforementioned assessment parameters based on the confusion matrix is given as follows:

$$\mathrm{ACC} = \frac{\mathrm{TP} + \mathrm{TN}}{\mathrm{TP} + \mathrm{TN} + \mathrm{FP} + \mathrm{FN}}$$

$$\mathrm{SENS} = \frac{\mathrm{TP}}{\mathrm{TP} + \mathrm{FN}}$$

$$\text{SPEC} = \frac{\text{TN}}{\text{TN} + \text{FP}}$$

$$\text{PREC} = \frac{\text{TP}}{\text{TP} + \text{FP}}$$

$$\text{ERR} = \frac{\text{FP} + \text{FN}}{\text{TP} + \text{TN} + \text{FP} + \text{FN}}$$

$$\text{MCC} = \frac{(\text{TP} * \text{TN}) - (\text{FP} * \text{FN})}{\sqrt{(\text{TP} + \text{FN}) * (\text{TP} + \text{FP}) * (\text{TN} + \text{FP}) * (\text{TN} + \text{FN})}}$$

1.6 Study Participants

In the present investigation, a total number of 50 archers were enrolled. The archers comprised of 37 males and 13 senior female youth who are in the age range of 13–20 with a mean and standard deviation of $M = 17.0$ and $\text{SD} = 0.56$, respectively. The archers were convened from a myriad of archery programmes in Malaysia. Normality test was conducted through a Shapiro–Wilk test, and the archers were found to be normality distributed with respect to the study variables.

1.7 Archery Shooting Test

Archery shooting test was set up for 50-m distance range in such a way that it mimics the standard international archery shooting competition. A total of 10 arrow shots were permitted to the archers; four shots were given as trials before recording the last six shots. It is worth noting that prior to any tests in the present study, informed consent was obtained in conformity with the guidelines of the Ethical Review Board of the Universiti Sultan Zainal Abidin (04-04/T-01/Jid 2.).

References

1. D.B. Jacobs, J.L. Park, Winning the team event at major international archery competitions. Proc. Inst. Mech. Eng. Part P J. Sport. Eng. Technol. **231**, 355–361 (2017)
2. R.M. Musa, M.R. Abdullah, A.B.H.M. Maliki, N.A. Kosni, S.M. Mat-Rasid, A. Adnan, H. Juahir, Supervised pattern recognition of archers' relative psychological coping skills as a component for a better archery performance. J. Fundam. Appl. Sci. **10**, 467–484 (2018)
3. W.D. McArdle, F.I. Katch, V.L. Katch, *Exercise Physiology: Nutrition, Energy, and Human Performance*. Lippincott Williams & Wilkins (2010)

4. S. Kagan, Z. Koruc, G. Latifoglu, Comparison of psychological and physiological changes of the anxiety in various sports. Rev. Cercet. SI Interv. Soc. **56**, 44–56 (2017)
5. Thakare V., *Comparative Study of Peak Expiratory Flow Rate of Archery Players Participated in All India Inter University Archery Competition* (2015)
6. A.H.K. Montoye, M. Begum, Z. Henning, K.A. Pfeiffer, Comparison of linear and non-linear models for predicting energy expenditure from raw accelerometer data. Physiol. Meas. **38**, 343–357 (2017)
7. K. Ellis, J. Kerr, S. Godbole, G. Lanckriet, D. Wing, S. Marshall, A random forest classifier for the prediction of energy expenditure and type of physical activity from wrist and hip accelerometers. Physiol. Meas. **35**, 2191–2203 (2014)
8. B. Barshan, M.C. Yuksek, Recognizing daily and sports activities in two open source machine learning environments using body-worn sensor units. Comput. J. **57**, 1649–1667 (2014)
9. S.G. Trost, Y. Zheng, W.-K. Wong, Machine learning for activity recognition: hip versus wrist data. Physiol. Meas. **35**, 2183–2189 (2014)
10. A. Joseph, N.E. Fenton, M. Neil, Predicting football results using Bayesian nets and other machine learning techniques. Knowl.-Based Syst. **19**, 544–553 (2006)
11. C. Peace, E. Okechukwu, An improved prediction system for football a match result. IOSR J. Eng. **04**, 2250–3021 (2014)
12. C.P. Igiri, Support vector machine-based prediction system for a football match result **17**, 21–26
13. N. Razali, A. Mustapha, F.A. Yatim, R. Ab Aziz, Predicting football matches results using bayesian networks for english premier league (EPL). IOP Conf. Ser. Mater. Sci. Eng. **226**, 012099 (2017)
14. Z. Ivankovi, M. Rackovi, B. Markoski, D. Radosav, M. Ivkovi, Analysis of basketball games using neural networks. Comput. Intell. Informatics (CINTI), 2010 11th Int. Symp. 251–256 (2010)
15. A. Maszczyk, A. Gołaś, P. Pietraszewski, R. Roczniok, A. Zając, A. Stanula, Application of neural and regression models in sports results prediction. Procedia—Soc. Behav. Sci. **117**, 482–487 (2014)
16. J. Edelmann-nusser, A. Hohmann, B. Henneberg, Modeling and prediction of competitive performance in swimming upon neural networks. Eur. J. Sport Sci. **2**, 1–10 (2002)
17. B.M. Collins, J.M. Lydersen, D.L. Fry, K. Wilkin, T. Moody, S.L. Stephens, Variability in vegetation and surface fuels across mixed-conifer-dominated landscapes with over 40 years of natural fire. For. Ecol. Manage. **381**, 74–83 (2016)
18. J. Han, J. Pei, M. Kamber, *Data Mining: Concepts and Techniques* (Elsevier, 2011)
19. P. Cunningham, S.J. Delany, k-nearest neighbour classifiers. Mult. Classif. Syst. **34**, 1–17 (2007)
20. P. Horton, K. Nakai, Better prediction of protein cellular localization sites with the k nearest neighbors classifier. Proceedings. Int. Conf. Intell. Syst. Mol. Biol. **5**, 147–52 (1997)
21. R. Mehrotra, A. Sharma, Conditional resampling of hydrologic time series using multiple predictor variables: a K-nearest neighbour approach. Adv. Water Resour. **29**, 987–999 (2006)
22. D.A. Adeniyi, Z. Wei, Y. Yongquan, Automated web usage data mining and recommendation system using K-nearest neighbor (KNN) classification method. Appl. Comput. Informatics. **12**, 90–108 (2016)
23. H. Holmström, M. Nilsson, G. Ståhl, Simultaneous estimations of forest parameters using aerial photograph interpreted data and the *k* nearest neighbour method. Scand. J. For. Res. **16**, 67–78 (2001)
24. C. Cortes, V. Vapnik, Support-vector networks. Mach. Learn. **20**, 273–297 (1995)
25. M. Negnevitsky, *Artificial Intelligence: a Guide to Intelligent Systems* (Pearson Education, 2005)
26. I.M. Yusri, A.P.P. Abdul Majeed, R. Mamat, M.F. Ghazali, O.I. Awad, W.H. Azmi, A review on the application of response surface method and artificial neural network in engine performance and exhaust emissions characteristics in alternative fuel. Renew. Sustain. Energy Rev. **90** (2018)
27. A.P. Worth, M.T. Cronin, The use of discriminant analysis, logistic regression and classification tree analysis in the development of classification models for human health effects. J. Mol. Struct. THEOCHEM. **622**, 97–111 (2003)

28. S.C. Lemon, J. Roy, M.A. Clark, P.D. Friedmann, W. Rakowski, Classification and regression tree analysis in public health: methodological review and comparison with logistic regression. Ann. Behav. Med. **26**, 172–181 (2003)
29. M. Schumacher, R. Roßner, W. Vach, Neural networks and logistic regression: part I. Comput. Stat. Data Anal. **21**, 661–682 (1996)
30. W. Vach, R. Roflner, M. Schumacher, *Neural Networks and Logistic Regression: Part II* (1996)

Chapter 2
Bio-Physiological Indicators in Evaluating Archery Performance

2.1 Overview

During archery performance, an archer is required to shoot an arrow to a target with a high degree of accuracy. To achieve a high degree of accuracy, several bio-physiological elements must be regulated both during the stance and the aiming stages. It has been reported that in archery performance, an archer is expected to regulate his/her breathing pattern, maintains a steady flow of blood to the working muscles (not to be under extreme tension), as well as has an optimum level of energy [1, 2]. Moreover, evidence has demonstrated that a higher degree of attention is needed for successful performance in archery. A greater level of attention could be achieved when proper consideration is devoted to certain cardiac indicators which constitute of heart rate, blood pressure level as well as breathing pattern [3]. The aforesaid bio-physiological parameters could influence archery performance and are therefore considered essential in defining the outcome of performance in the sport. The present investigation aims to identify the association of specific bio-physiological parameters with the archery performance and to classify high- and low-performance archers based on their performance in the selected bio-physiological parameters, viz. resting heart rate, resting, respiratory rate, calories intake, resting systolic and diastolic blood pressure.

The following procedures were taken in order to achieve the objective of the investigation:

Dietary Intake Evaluation: The dietary intake of the archers was evaluated by means of a three-day dietary recall. A self-administered form was devised for this purpose. The form contains sections and columns for the information on the type of food taken, the serving size, as well as the time the food is consumed. In order to ensure the reliability of the evaluation, the archers were trained on how to measure and fill the form each time the meal or drink is taken. The dietary intake for two-week days and a weekend was recorded and analysed via a nutritionist pro analysis software.

R. Muazu Musa et al., *Machine Learning in Sports*, SpringerBriefs in Applied Sciences and Technology, https://doi.org/10.1007/978-981-13-2592-2_2

Resting Heart Rate, Diastolic and Systolic Blood Pressure Measurement: The measurements of the resting heart rate, diastolic as well as systolic blood pressure were carried out early in the morning before any form of exercises or warm up. Omron (HEM-7120) blood pressure and heart rate measurement device was used to measure the aforesaid parameters. The device has previously been reported as a reliable tool for measuring the interrelated cardiac parameters [4]. The parameters were measured whilst the archers were in a sitting position comfortably on a chair.

Resting Respiratory Rate Assessment: A stopwatch was utilised to record the resting respiratory rate of the archers for a period of 1 min. The test was implemented whilst the archers were sitting upright on a chair. The breaths of the archers were recorded manually by a therapist. The evaluation was carried out three times, and the maximum value was used as suggested by the previous researchers [5].

2.2 Clustering

The k-means clustering algorithm was utilised in the present study. The number of clusters is predetermined as two to discriminate against LPA and HPA. The clustering is governed by the bio-physiological parameters (features) selected in the study as well as the archery shooting score. The classes are assigned based on the considered features. The cluster analysis was performed using Orange v3.11.

2.3 Classification

In this study, logistic regression and a different variety of k-NN algorithms, namely fine, medium, coarse, cosine, cubic and weighted, are investigated with regard to its efficacy in providing the better classification on the clustered HPA and LPA data. The classification accuracy of the evaluated models was obtained through the fivefold cross-validation technique. Through this cross-validation technique, the data is divided into five sets, in which four of the five sets of the data are used for training and the remaining one set is used as the validation set. The performance of the classifiers is evaluated with respect to different classification performance measures as indicated in the subsequent section. The classification analysis was carried out using MATLAB (2016a).

2.4 Results and Discussion

Figure 2.1 illustrates the box plot analysis of the clustered data via the k-means clustering algorithm. It is noticeable from Fig. 2.1a, c and d that the score of the measured variables is considerably high. The HPA class is attributed to these variables. Conversely, Fig. 2.1b, e and f demonstrated a higher score associated with the LPA class. It is evident from the box plot analysis that the HPA class has a higher calorie intake and respiratory rate; however, a lower resting heart rate, resting systolic and diastolic blood pressure are demonstrated by the aforesaid class.

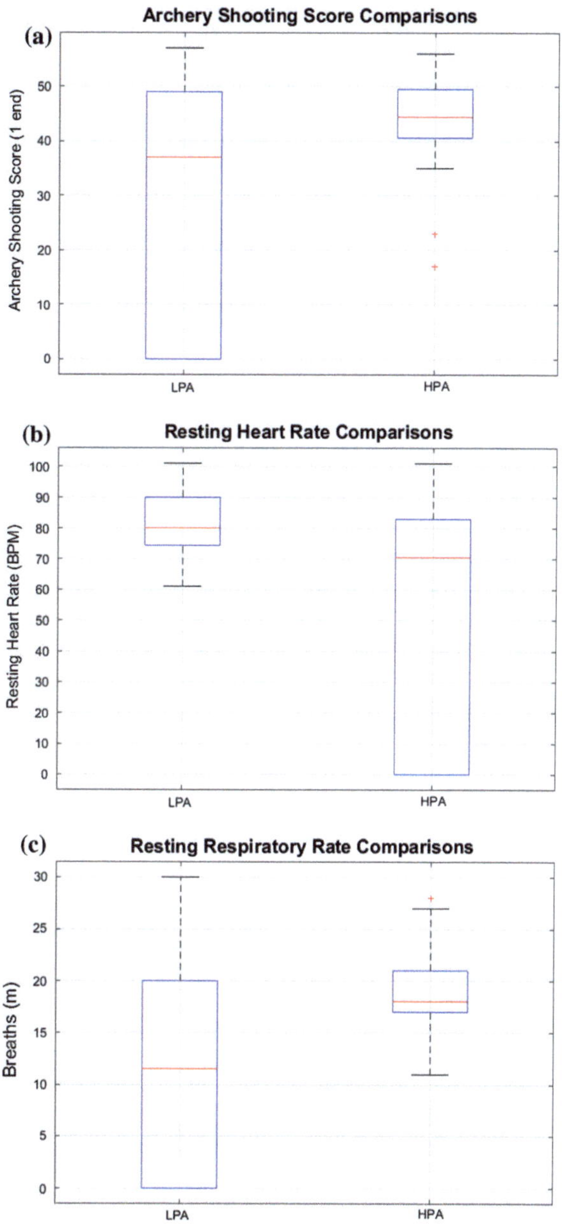

Fig. 2.1 Bio-physiological box plot classes. **a** Archery shooting score, **b** resting heart rate, **c** resting respiratory rate, **d** calories intake, **e** resting systolic blood pressure, **f** resting diastolic blood pressure

Fig. 2.1 (continued)

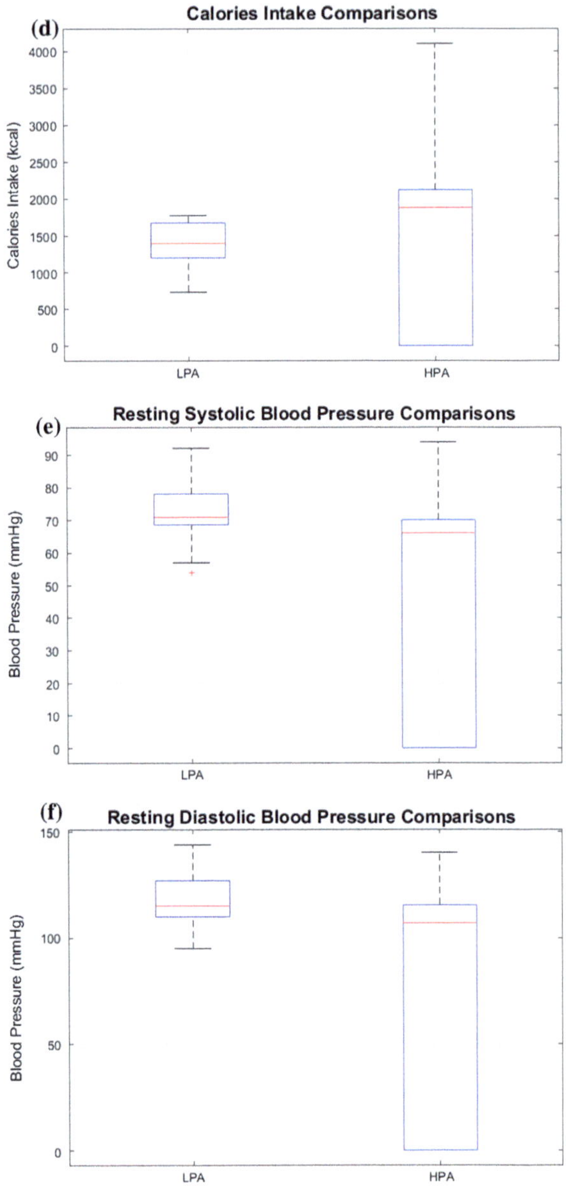

Table 2.1 Model evaluation

Algorithms	ACC (%)	SENS (%)	SPEC (%)	PREC (%)	ERR (%)	MCC
LR	96.00	94.44	96.88	94.44	4.00	0.9132
Fine *k*-NN	74.00	61.11	81.25	64.71	26.00	0.4292
Medium *k*-NN	82.00	66.67	90.63	80.00	18.00	0.6001
Coarse *k*-NN	64.00	0.00	100.00	N/A	36.00	N/A
Cosine *k*-NN	76.00	90.32	68.42	82.35	18.00	0.6113
Cubic *k*-NN	82.00	61.11	93.75	84.62	18.00	0.6003
Weighted *k*-NN	84.00	61.11	96.88	91.67	16.00	0.6517

It is apparent from Table 2.1 that the logistic regression classifier is a better classifier against all the variation of the k-NN algorithms based on all the classification performance metrics. Amongst the k-NN algorithms, the weighted k-NN algorithm seems to provide a reasonable good classification accuracy with an accuracy of 84%. Although the classification accuracy appears to be relatively good, it misclassified seven HPA as LPA; this is evident from the confusion matrix depicted in Fig. 2.2 as well as the from the SENS metric which considers the true positive rate or the correct positive prediction. Moreover, the MCC metric also indicates a lower correlation in comparison to the LR algorithm. By far the worst-performing classifier is the coarse variation of the k-NN algorithm that provides a classification accuracy of 64%. This particular classifier misclassifies all the HPA to LPA; this is evident from the SENS metric, which in turn, resulted in an uncorrelated MCC.

The results from the investigations of the bio-physiological variables in the evaluation of the archery performance have demonstrated that high performance in the sport of archery is associated with possession of a certain level of bio-physiological indicators. It is apparent from the present findings that high-performance archers are recognised with higher respiratory rate, calories intake and a considerable lower heart rate as well as blood pressure. On the other hand, low-performance archers are ascribed with lower calories intake, lower respiratory rate and a higher blood pressure coupled with a higher heart rate (see Fig. 2.1).

Breathing technique has been previously reported to play a significant role in determining aiming stability as well as shooting consistency [1]. The authors have also reported a significant association between the breathing technique and shooting performance of elite and non-elite archers. In another study, the capacity of lung function has been discovered to be essential in the performance of archery. The researchers inferred that elite archers are characterised with higher respiratory frequency as compared to the non-elite archers [6]. High respiratory frequency enables the archer to control the arousal level incur during a competitive match. Moreover, the ability of the archer to control the high level of arousal is found to increase shooting consistency during archery performance [3].

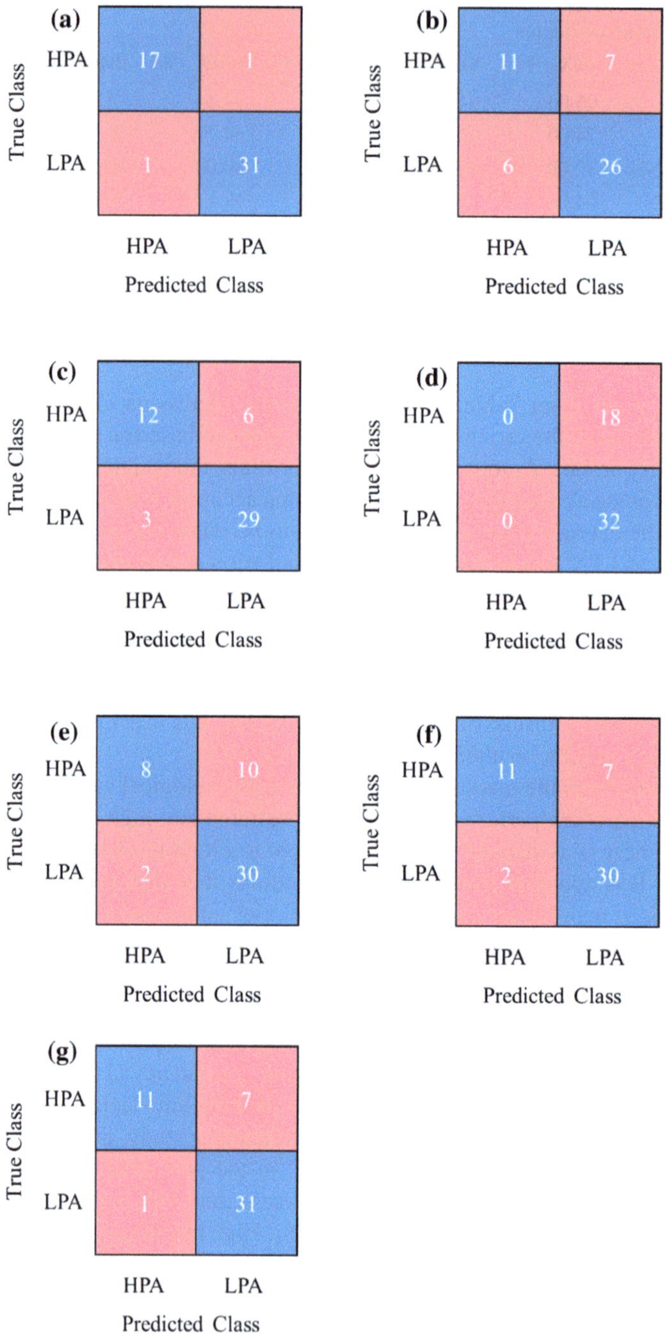

Fig. 2.2 Confusion matrix. **a** LR model, **b** fine k-NN model, **c** medium k-NN model, **d** coarse k-NN model, **e** cosine k-NN model, **f** cubic k-NN model, **g** weighted k-NN model

In another perspective, dietary intake is found to be non-trivial to archery performance. It has been revealed that nutrition is necessary for optimum athletic performance because it helps to provide an athlete with energy during the recruitment of the primary muscles of the body for performing sporting activity [7]. Archery is an energy-demanding sport in which the archers spend a considerable amount of time shooting on the range both during practice and competition. It is therefore vital to keep the working muscle powered at the time of the shooting. The archers who are less energetic as a result of lack of eating or poor dietary intake are likely to become frail and experience a performance setback [8].

A greater level of blood pressure as well as heart rate is found to be associated with low-performance archers. The higher the degree of both systolic and diastolic blood pressure coupled with a maximum heart rate could be a sign of extreme anxiety which in turn reciprocate to a high degree of arousal. The sport of archery is highly competitive in which a performance outcome of either a winner or loser could be determined by a score of a single arrow shot [9]. It has also been documented that a typical archery competition range is crowded with a high pitch noise that could lead a competitor to a state of apprehension through an increase in a higher heart rate activity as well as an increase in blood level [10]. Hence, the ability of an archer to control as well as lessen the arousal level during the competition is essential for physiological stabilisation which could lead to a shooting precision and consistency.

2.5 Summary

The evaluation of archery performance with regard to the bio-physiological indicators has been carried out in this investigation. The findings from the present investigation have demonstrated that certain bio-physiological markers, namely calories intake, resting respiration, heart rate, systolic as well as diastolic blood pressure could influence archery performance. It was established from the investigation that a high calories intake, respiratory rate and a lower level of heart rate as well as lower corresponding blood pressure are positively linked with a higher archery shooting score. Moreover, it is shown from the present investigation that the LR model is able to provide a reasonably good prediction of LPA and HPA based on the selected features.

References

1. M.N. Mohamed, W.M.N. Wan Norman, A. Linoby, M.H. Sariman, M.Z. Mohd Azam, Breathing pattern influence to the shooting performance, in *Proceedings of the International Colloquium on Sports Science, Exercise, Engineering and Technology* 2014 (ICoSSEET 2014) (Springer Singapore, Singapore), pp. 321–333 (2014)
2. R.M. Musa, M.R. Abdullah, A.B.H.M. Maliki, N.A. Kosni, S.M. Mat-Rasid, A. Adnan, H. Juahir, Supervised pattern recognition of archers' relative psychological coping skills as a

component for a better archery performance. J. Fundam. Appl. Sci. **10**, 467–484 (2018)
3. C. Robazza, L. Bortoli, V. Nougier, Physiological arousal and performance in elite archers: a field study. Eur. Psychol. **3**, 263–270 (1998)
4. M.R. Abdullah, V. Eswaramoorthi, R.M. Musa, M. Maliki, A.B. Husin, N.A. Kosni, M. Haque, The effectiveness of aerobic exercises at difference intensities of managing blood pressure in essential hypertensive information technology officers. J. Young Pharm. **8** (2016)
5. D. Han, M. Ha, Effect of pelvic floor muscle exercises on pulmonary function. J. Phys. Ther. Sci. **27**, 3233–3235 (2015)
6. Thakare, V., *Comparative Study of Peak Expiratory Flow Rate of Archery Players Participated in All INDIA Inter University Archery Competition* (2015)
7. Z. Taha, R.M. Musa, A.P.P.A. Majeed, M.R. Abdullah, M. AmirulAbdullah, M.H.A. Hassan, Z. Khalil, The employment of support vector machine to classify high and low performance archers based on bio-physiological variables. IOP Conf. Ser. Mater. Sci. Eng. **342**, 12020 (2018)
8. K. Chaudhary, I. Sukhwal, Nutrition for optimal sports performance-a review. Int. J. Recent Sci. Res. **7**, 9988–9992 (2016)
9. Z. Taha, R.M. Musa, A.P.P. Abdul Majeed, M.M. Alim, M.R. Abdullah, The identification of high potential archers based on fitness and motor ability variables: a support vector machine approach. Hum. Mov. Sci. (2018)
10. Z. Taha, R.M. Musa, A.P.P.A. Majeed, M.R. Abdullah, M.M. Alim, A.F.A. Nasir, The application of k-nearest neighbour in the identification of high potential archers based on relative psychological coping skills variables. IOP Conf. Ser. Mater. Sci. Eng. **342**, 12019 (2018)

Chapter 3
Psychological Variables in Ascertaining Potential Archers

3.1 Overview

A successful archery performance to a larger extent depends on several performance factors. The very nature of the archery sport as a static and precision sport entails adequate preparation both during and before a competition. A standard archery competition is identified with background noises, tension, anxiety as well as excitements, for this reason, a proper mental adjustment and focus are necessary for success in the sport [1]. Research has shown that some psychological elements such as the feeling of worry, stress and lack of focus hamper athletic performance particularly in the sport of archery in which mental toughness is one of the keys to its successful enactment [2]. However, it is worth noting that the requirement of a specific psychological skill varies from one sport to another. The peculiarity of certain sport places a unique demand of psychological preparation for the athletes partaking in the sport. It is therefore pertinent to evaluate the psychological elements necessary for a particular sport.

Athletic coping skills inventory (ACSI) is an instrument developed by the preceding researchers to assess the ability of an athlete in dealing with various coping elements [3]. In this investigation, ACSI is applied to ascertain potential archers. The instrument is considered appropriate for usage in the present study due to the nature of archery as an individual sport in which the archer is expected to compete alone against an opponent.

Athletic Coping Skills Evaluation: The ACSI measures seven coping skill constructs, viz. coachability, concentration, confidence and achievement motivation, goal setting and mental preparation, peaking under pressure as well as freedom from worry. The coachability construct evaluates the ability of an athlete to be open-minded with different coaching styles as well as accept criticism without being upset. Concentration assesses the athlete capability to be focused and ignore external influences that could distract attention. The confidence and achievement motivation explore the ability of an athlete to remain motivated and always delivers the best possible per-

R. Muazu Musa et al., *Machine Learning in Sports*, SpringerBriefs in Applied Sciences and Technology, https://doi.org/10.1007/978-981-13-2592-2_3

formance during both practice and competition. Goal setting and mental preparation measure self-directed goals as well as plans an athlete sets with the strategic guidelines and game plans to achieve the predetermined set goals. Peaking under pressure endeavours to assess the capacity of an athlete to do well under pressure, to accept challenge and to avoid being intimidated by fear, anxiety or stress. Finally, freedom from worry evaluates the tenacity of an athlete to disregard any worry, pressure or concern about performing poorly.

The ACSI scores of the archers were determined prior to the shooting assessment. It is worth to mention that the instrument was pretested to further validate the reliability of the items for the participants under study. Acceptable values ranging from 0.67 to 0.82 were retrieved affirming acceptable reliability as well as internal consistency of the constructs.

3.2 Clustering

The clustering method applied for the present investigation is the k-means clustering algorithm. The number of clusters is determined by the silhouette score governed by the Euclidean distance metric, and from the data, the number of k ascertained was found to be two which suggests that the data can evidently be clustered into HPA and LPA. Figure 3.1 depicts the silhouette score of the present data. Orange v3.11 was used for the clustering analysis.

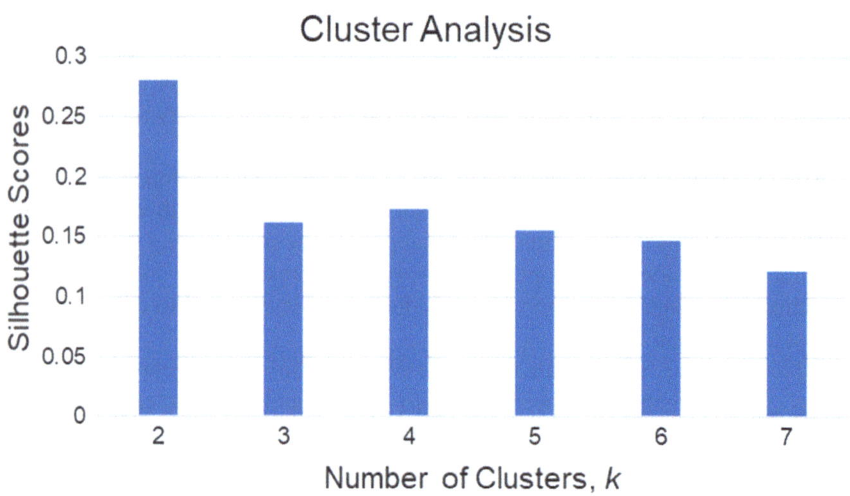

Fig. 3.1 Cluster analysis by means of k-means

3.3 Classification

In this study, Artificial Neural Network (ANN) is employed to classify the performance classes of the archers. Two different network architectures are investigated with respect to its ability to classify well the archers' performance. The difference between the architectures lies with the activation function employed at the hidden layer, namely tansig and logsig activation functions. The number of neurones at the hidden layer is ascertained to be 26 based on the following equation suggested by Jinchuan and Xinzhe:

$$N_h = \frac{N_{\text{in}} + \sqrt{N_p}}{L}$$

where N_{h} is the number of input neurones, i.e. the physiological parameters selected, N_{p} is the number of input sample whilst L is the number of the hidden layers which in this study is one.

A fivefold cross-validation technique is utilised in order to obtain the classification accuracy of the developed ANN models. The data (50 observations) is partitioned into five sets where four sets of the data are used for training whilst one set is used for the validation set. The ANN analysis was carried out using Orange v3.11.

3.4 Results and Discussion

The comparison of the box plot scores for the psychological variables is shown in Fig. 3.2. It is clear from the plots that the mean performances of the HPA class are much higher in comparison to the LPA class except for the freedom from worry variable as shown in Fig. 3.2g. The analysis has indicated that the HPA is psychologically well prepared.

Two ANN models with a network topology of 8-26-1 with different activation functions were evaluated. The confusion matrix for both models is depicted in Fig. 3.3. Based on the classification accuracy given in Table 3.1 which is derived from the confusion matrix, it is evident that the tansig-based ANN model is a better model with a classification accuracy of 94%. The model is able to provide a relatively balanced classification of both LPA and HPA as demonstrated in Fig. 3.3a. Conversely, although the logsig-based ANN model provided no misclassification on the HPA as demonstrated by the sensitivity evaluation, nonetheless, eight athletes that fall under the category of LPA is misclassified as HPA, this in turn reduces the classification accuracy of the model to only 84%. Although 84% appears to be a reasonably high classification accuracy, it is noteworthy to mention that the model could be misleading for the identification of potential archers by coaches since the classification for both LPA and HPA is non-trivial. Therefore, the insight provided by

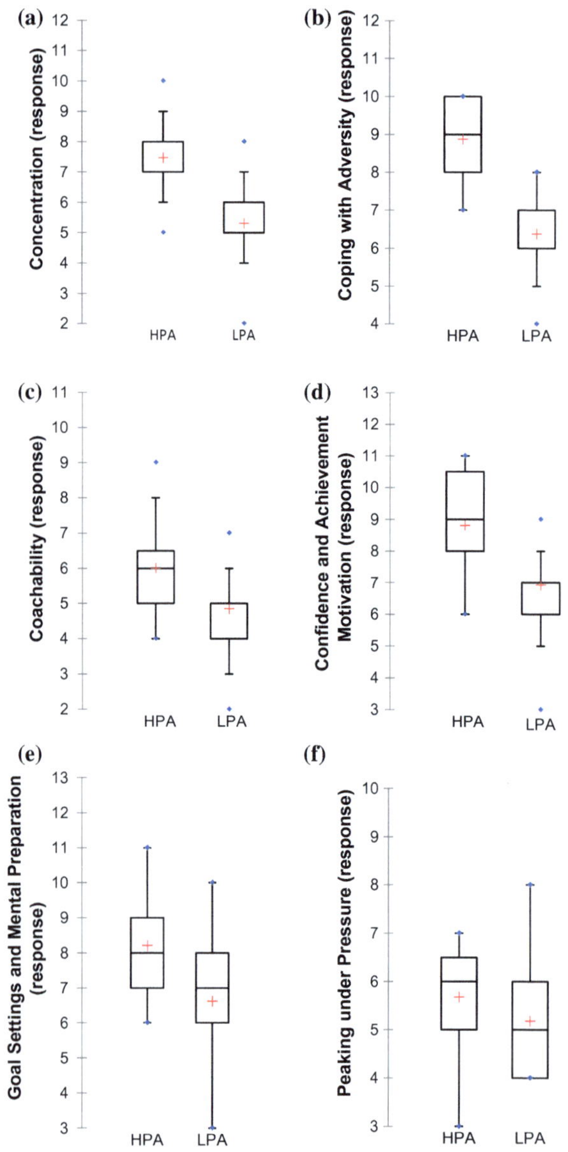

Fig. 3.2 Psychological box plot classes. **a** Concentration, **b** coping with adversity, **c** coachability, **d** confidence and achievement motivation, **e** goal settings and mental preparation, **f** peaking under pressure, **g** freedom from worry, **h** archery shooting score

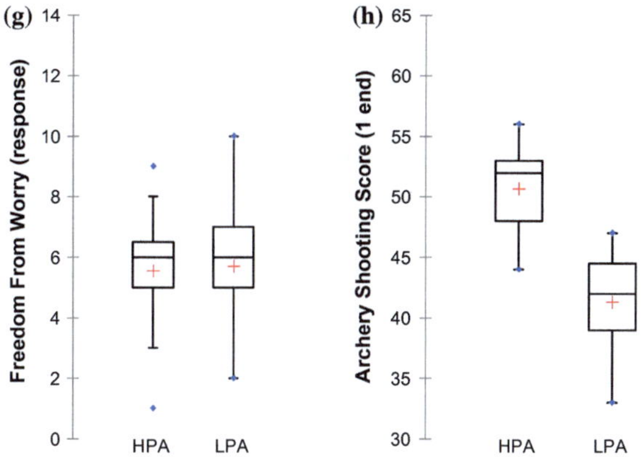

Fig. 3.2 (continued)

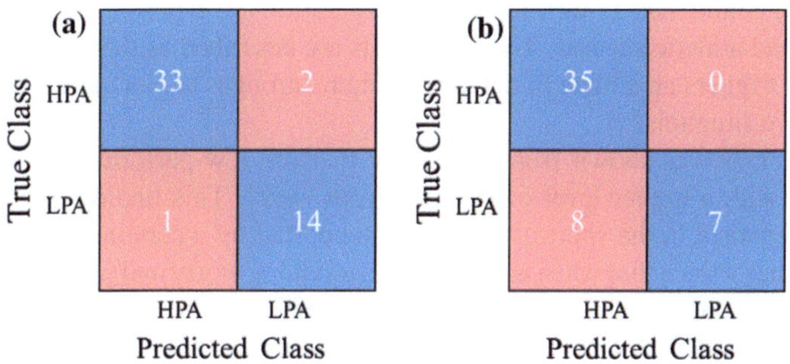

Fig. 3.3 Confusion matrix. **a** Tansig-based ANN model and **b** logsig-based ANN model

Table 3.1 Model evaluation

Algorithms	ACC (%)	SENS (%)	SPEC (%)	PREC (%)	ERR (%)	MCC
ANN-tansig	94.00	94.29	93.33	97.06	6.00	0.8608
ANN-logsig	84.00	100.00	46.67	12.00	16.00	0.6163

the SENS and SPEC evaluation matrix in this case is invaluable. Moreover, the strong relationship provided by the MCC and low error rate provided by the tansig-based ANN model further demonstrate the efficacy of the model.

This study is carried out to identify potential archers by means of a number of psychological coping skill variables. ACSI was used as the instrument of data collection in which the archers' scores of the instruments and their archery shooting score are determined. It was discovered that the athletes who recorded a higher coping skill ability have a corresponding greater archery shooting score.

The essential findings from this study have revealed that the predetermined psychological coping skill variables are able to separate the high- and low-performance archers. It was established that the combination of ANN-tansig model is able to provide a rather reasonable classification for the discrimination of the high- and low-performance archers with an excellent accuracy of 94% amongst the other model's performance metrics. The findings have generally underpinned the inevitability of psychological coping skills in the sport of archery.

The sport of archery is regarded as a close skill in which a competitor is required to compete on a face-to-face shooting pattern which poses the need for the archers to be mentally tough for a better delivery of performance. The high-performance archers in this study are attributed to a greater degree of concentration, confidence and achievement motivation, goal setting and mental preparation, coachability as well as peaking under pressure as shown in Fig. 3.2. It has been documented that psychological elements such as confidence, self-motivation and anxiety control are the vital elements towards achieving success in a sporting activity [4]. Moreover, concentration, confidence as well as mental preparation are reported to be the driving force behind athletic success. These elements are described as the essential prerequisite that triggers an athlete in defeating a more stronger opponent during a highly competitive situation [5].

However, it was shown from this study that the low-performance archers are identified with a higher level of freedom from worry. This finding illustrates that low performance in the sport of archery is supported by a certain degree of worry. Evidence has shown that when an athlete is in a state of worry and pay more attention to what people might think of the outcome of the match rather than the match at hand, the athlete might likely suffer setback and an eventual failure [6]. It is therefore important for archers to be free from worry so as to deliver their best performance.

3.5 Summary

The present study has explored the association of athletic psychological coping skills with the performance of archery sport. It was demonstrated from the study that the possession of certain psychological coping skills, namely coping with adversity, coachability, peaking under pressure, concentration, confidence and achievement motivation, goal setting and mental preparation could aid an archer to endure the pressure and cope during the competition which in turn could result in high archery performance. However, a higher level of freedom from worry could hamper successful performance in archery sport. The study has further shown that high-performance archer could be identified based on their ability to cope with the psychological issues arising during a competitive archery match. Moreover, with regard to the machine learning aspect, it was demonstrated that the tansig-based ANN model could provide the better classification of HPA and LPA.

References

1. H.A. Aysan, The effects of archery as a sports branch on coping with stress. Stud. Ethno-Medicine. **10**, 39–43 (2016)
2. R.S. Lazarus, How emotions influence performance in competitive sports. Sport Psychol. **14**, 229–252 (2000)
3. R.E. Smith, R.W. Schutz, F.L. Smoll, J.T. Ptacek, Development and validation of a multidimensional measure of sport-specific psychological skills: the athletic coping skills inventory-28. J. Sport Exerc. Psychol. **17**, 379–398 (1995)
4. M.R. Abdullah, R.M. Musa, A.B.H.M.B. Maliki, Kosni, N.A., Suppiah, P.K.: Role of psychological factors on the performance of elite soccer players. J. Phys. Educ. Sport. **16**, 170 (2016)
5. M.R. Abdullah, A. Maliki, R.M. Musa, N.A. Kosni, H. Juahir, S.B. Mohamed, Identification and comparative analysis of essential performance indicators in two levels of soccer expertise. Int. J. Adv. Sci. Eng. Inf. Technol. **7**, 305–314 (2017)
6. Z. Taha, R.M. Musa, A.P.P.A. Majeed, M.R. Abdullah, M. Aizzat Zakaria, M.M. Alim, J.A.M. Jizat, M.F. Ibrahim, The identification of high potential archers based on relative psychological coping skills variables: a support vector machine approach. IOP Conf. Ser. Mater. Sci. Eng. **319**, 12027 (2018)

Chapter 4
Anthropometry Correlation Towards Archery Performance

4.1 Overview

Anthropometry refers to the attributes that describe the physical and the anatomical features of an individual. It involves the composition, size and shape of the human body. In a sporting activity, some evidence has suggested that anthropometry variables such as the size, shape as well as the composition of the body could determine the strength of an individual which in turn contribute towards performing motor activity [1]. Moreover, a significant positive relationship between strength and physical characteristics in sports has also been documented [2]. The study has shown that stronger athletes could be identified with a peculiar body size, composition and shape who also tend to perform better during the execution of motor activity as compared to their counterparts.

Whilst it is important to stress that the body physique of an athlete may not necessarily predict performance in certain sports, it is worth to mention that sporting activity that is connected to resistance actions of pulling and lifting as in the case of archery could require a specific body composition. In spite of the inadequacy of the relevant data on anthropometry and its association to archery performance, the previous literature has demonstrated that the body composition and the shape of the Xibo ethnic archers differ from one another [3]. The variation of the body shapes and composition of the archers were reported to have an influence on their overall archery performance. However, the study mainly focused on the particular body attributes that might be required for archery performance with regard to male and female. It is, therefore, needful to investigate the possible correlation of anthropometry characteristics towards archery performance where the present investigation is built upon.

Anthropometry measurements: The standard techniques for the measurement of the anthropometric attributes were used in this investigation. The standing height was measured using a wall-mounted wooden stadiometer, whilst the weight was determined using an electronic digital scale. The arm span, calf, hip, thigh as well as

R. Muazu Musa et al., *Machine Learning in Sports*, SpringerBriefs in Applied Sciences and Technology, https://doi.org/10.1007/978-981-13-2592-2_4

abdominal circumference were measured using a measuring tape. The ISAK protocol for the measurement of anthropometry-related variables was utilised in this study [4]. The archery shooting score of the archers was determined prior to the anthropometric measurements. A total of ten arrow shots were permitted. The first four shots were given as a trial, whilst the final six shots were recorded for the statistical analysis.

4.2 Clustering

The present study utilises the k-means clustering algorithm. The number of clusters is determined *a priori* into two clusters, namely HPA and LPA. These clusters are determined by the minimisation of the intra-cluster variance attributed by the anthropometry indicators as well as the archery shooting score evaluated in this study. The Euclidean distance objective function is employed for this exercise. The cluster analysis was performed using Orange v3.11.

4.3 Classification

A different variation of SVM models with different kernel functions is evaluated, i.e. linear, quadratic, cubic, fine RBF, medium RBF as well as coarse RBF. These classifiers are evaluated based on its ability to classify the clustered HPA and LPA data. The fivefold cross-validation technique is employed in this study owing to the size of the available data. Different performance metrics are utilised in order to evaluate the ability of the classifiers to classify well the HPA and LPA, namely classification accuracy (CA), SENS, SPEC, PREC, ERR and MCC. The classification analysis was performed by means of MATLAB (2016a).

4.4 Results and Discussion

Figure 4.1 illustrates the box plots attained through the clustering analysis via k-means. It could be seen that the HPA is highly influenced by the archery shooting score and all corresponding anthropometry variables except for the calf circumference as shown in Fig. 4.1(d). Therefore, it is apparent from this analysis, and a considerable insight into the relevant anthropometric parameters that influence the performance on the sport of archery could be determined.

Table 4.1 shows the performance of the variation of the SVM models assessed in this investigation. It is obvious that the linear-based SVM model provides the highest classification accuracy as well as relatively good evaluation of other performance metrics. Although the SPEC is slightly low in comparison with the cubic and coarse RBF-based SVM algorithms, however, the linear-based model still

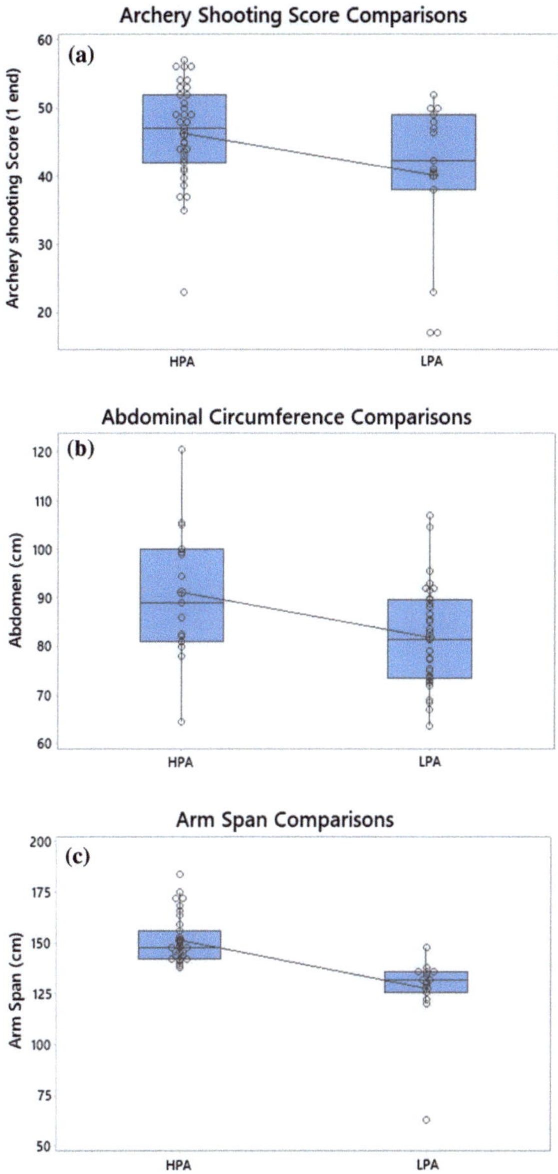

Fig. 4.1 Anthropometry box plot classes. **a** Archery shooting score, **b** abdominal circumference, **c** arm span, **d** calf circumference, **e** height, **f** hip circumference, **g** thigh circumference and **h** weight

provides prediction due to the nature of the selected variable. The worst performing SVM classifier is the fine RBF variation as it misclassifies all the LPA as HPA as demonstrated by the SPEC as well as the confusion matrix shown in Fig. 4.2.

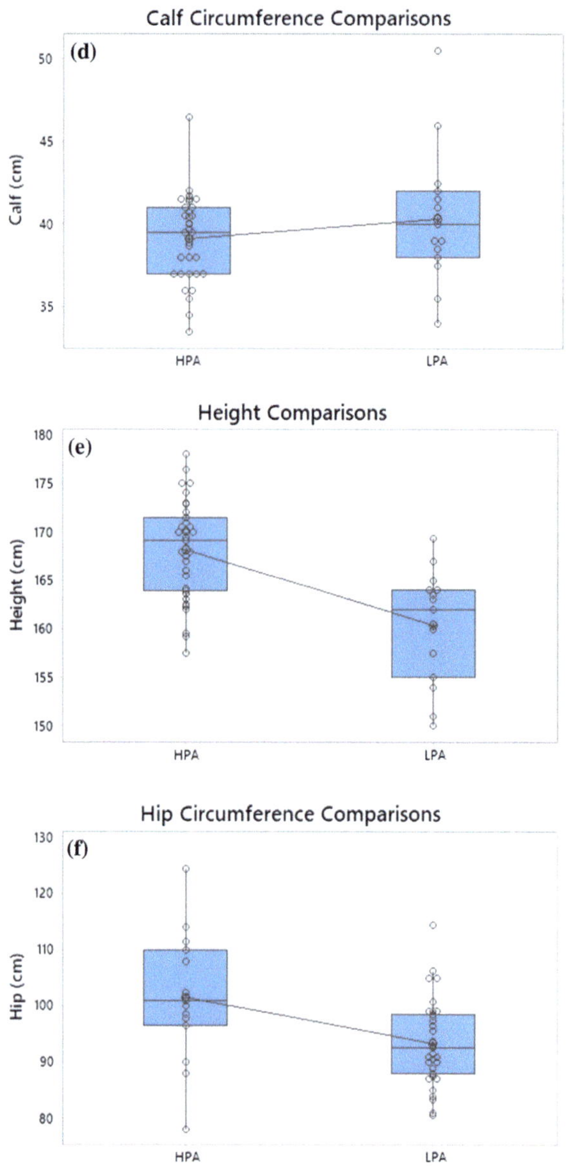

Fig. 4.1 (continued)

The present investigation endeavours to examine the correlation of the anthropometric variables with the performance of archery sport. The study is geared towards the identification of the suitable physical attributes relevant to the sport of archery and to also identify potential archers by virtue of their natural anthropometric characteristics. It is demonstrated in the study that successful performance in the sport

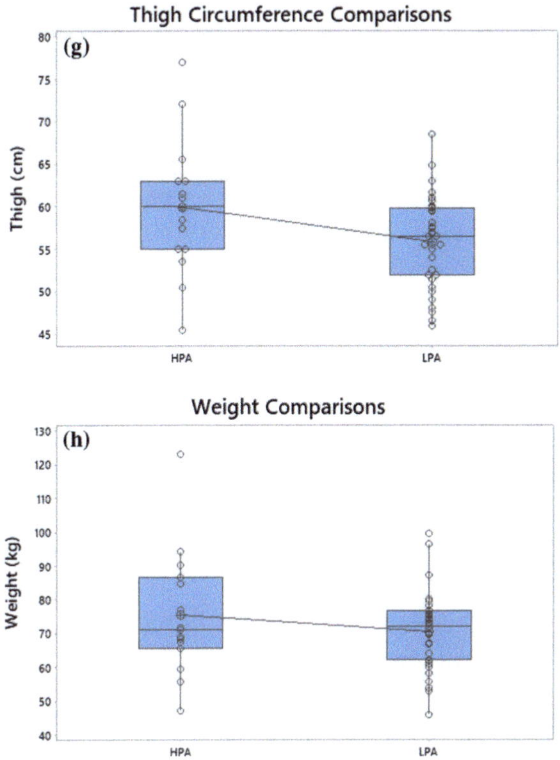

Fig. 4.1 (continued)

Table 4.1 Model evaluation

Algorithms	ACC (%)	SENS (%)	SPEC (%)	PREC (%)	ERR (%)	MCC
SVM-L	88.00	100.00	60.00	85.37	12.00	0.7157
SVM-Q	86.00	97.14	60.00	85.00	14.00	0.6547
SVM-C	84.00	91.43	66.67	86.49	16.00	0.6069
SVM-F-RBF	70.00	100.00	0.00	70.00	30.00	N/A
SVM-M-RBF	78.00	97.14	33.33	77.27	22.00	0.4298
SVM-C-RBF	72.00	100.00	66.67	71.43	28.00	0.2182

of archery could be heavily reliant upon the physical characteristics of the archer as evident from Fig. 4.1. It could similarly be observed from the figure that the HPA is characterised with a greater archery shooting score and all corresponding anthropometry variables except for the calf circumference. The findings have further illustrated that athletes with a bigger body size could perform the sport of archery better as opposed to the small-sized athletes.

Fig. 4.2 Confusion matrix. **a** SVM-L model, **b** SVM-Q model, **c** SVM-C model, **d** SVM-F-RBF model, **e** SVM-M-RBF model and **f** SVM-C-RBF model

Archery sport is considered as an energy-demanding sport. The archers are required to carry, lift and pull bow and arrow in the process of shooting whilst maintaining a static balance. In order to complete the aforesaid task, an enormous amount of energy is necessary. It has been reported from the previous study that athletes that are bigger in size are able to maintain a static balance and counter the effect of body sway during the performance of static physical activity [5]. The study has further shown that postural sway is linked with lanky athletes who often strive to maintain balance in a static position. It is worth to mention at this juncture that successful performance in archery could be significantly influenced by postural sway. For an archer to sustain precision and accuracy in the shooting competition, the archer needs to counter any form of postural sway to hit a predetermined target. Another study has also documented a positive relationship between successful archery performance and endomorphs body attributes [6]. The authors inferred that endomorphs body-typed athletes who are primarily related to pear-shaped, high body fat, bigger and rounded abdominal circumference as well as a broader thigh are able to achieve a maximum archery score in comparison with the shorter and smaller archers.

4.5 Summary

It has been identified in the present study that successful performance in the sport of archery could be influenced by the archer's physical attributes. It has been discovered that a bigger body size is positively correlated with a successful archery performance. The archers who are bigger in size could be able to sustain balance, counter the effect of postural sway as well as possess the energy required to carry out the demand of tasks inherent in performing archery sport.

References

1. S. Ostojic, S. Mazić, N. Dikic, S.M. Ostojic, S. Mazic, Profiling in basketball: physical and physiological characteristics of elite players. Artic. J. Strength Cond. Res. **20**, 740–744 (2006)
2. T.E. Ball, B.H. Massey, J.E. Misner, B.C. Mckeown, T.G. Lohman, The relative contribution of strength and physique to running and jumping performance of boys 7–11. J. Sports Med. Phys. Fitness **32**, 364–371 (1992)
3. Y. Wang, Y. Ma, Physique characteristics of ethnic archers based on R-cluster matrix (2015)
4. M.J. Marfell-Jones, A.D. Stewart, J.H. de Ridder, International standards for anthropometric assessment (2012)
5. H. Berenjian Tabrizi, A. Abbasi, H. Jahadian, Comparison of static and dynamic balance and its relationship with anthropometric characteristics in athletes of selected sports. J. Sport Sci. **6**, 33–36 (2014)
6. Z. Taha, M. Haque, R.M. Musa, M.R. Abdullah, A.B.H.M. Maliki, N. Alias, N.A., Kosni, Intelligent prediction of suitable physical characteristics toward archery performance using multivariate techniques. J. Glob. Pharma. Technol. **9** (2017)

Chapter 5
Psycho-Fitness Parameters in the Identification of High-Potential Archers

5.1 Overview

The sport of archery is multitasking in its nature. The amount of energy expends in the process of carrying the bow and arrow, and the constant movement whilst shooting and retrieving the arrow shot as well as the psychological needs to cope with the pressure, anxiety and tension during competition could place a high demand of physical fitness and psychological preparation to the archers. Archery sport takes a longer hour to complete both during training and competition. Therefore, the ability of an archer to be physically fit and psychologically tough could be beneficial to the accomplishment of a high archery score.

It has been previously reported in a number of studies that physical fitness could enable an archer to withstand the more extended hour of both practice and competition and as such contributes towards a successful performance in the sport [1, 2]. In another study, evidence was found supporting the needs of a psychological coping ability in performing the sport of archery [3].

Although all the studies mentioned above examined the requirements of the physical fitness and the psychological elements towards the attainment of a higher archery score, it is worth noting that the studies investigated the variables somewhat separately. It is, therefore, vital to look into the combination as well as the interactions of both the psychological coping skills and the physical fitness parameters (psycho-fitness) with respect to the archery performance. Thus, the present study intends to examine the interaction of the aforesaid parameters and consequently identifies high-potential archers with regard to their performance in the psycho-fitness parameters.

Psycho-fitness parameters assessments: The physical fitness parameters assessments which included hand grip, upper muscle strength, core muscle strength, static balance and vertical jump were conducted in accordance with the standard procedures for the fitness evaluation described by the previous researchers [4]. The athletic coping skills inventory (ACSI) was administered to the archers in order to determine the level of their psychological coping skills. The ACSI essentially measures the level of

R. Muazu Musa et al., *Machine Learning in Sports*, SpringerBriefs in Applied Sciences and Technology, https://doi.org/10.1007/978-981-13-2592-2_5

athletic coping ability of various psychological elements ranging from concentration, coachability, mental preparation and freedom from worry amongst others possessed by an athlete. The scores of both the fitness assessment and the psychological coping skills evaluation were recorded before the archery shooting test.

5.2 Clustering

The clustering of the psycho-fitness attributes with the archery shooting score was carried out through the k-means clustering algorithm via Orange v3.11 platform. The number of clusters is ascertained through the silhouette scores, and it was found that two clusters could discriminate well the HPA and LPA. Figure 5.1 illustrates the silhouette score that indicates that two clusterings are sufficient to provide a reasonable interpretation of the data clustered.

5.3 Classification

A variety of k-NN algorithms, viz. fine, medium, coarse, cosine, cubic and weighted, are examined with respect to its effectiveness in classifying the clustered HPA and LPA data. A fivefold cross-validation technique was utilised in the investigation. The k-NN models are evaluated with respect to different classification performance metrics, and the analysis was accomplished by means of MATLAB (2016a).

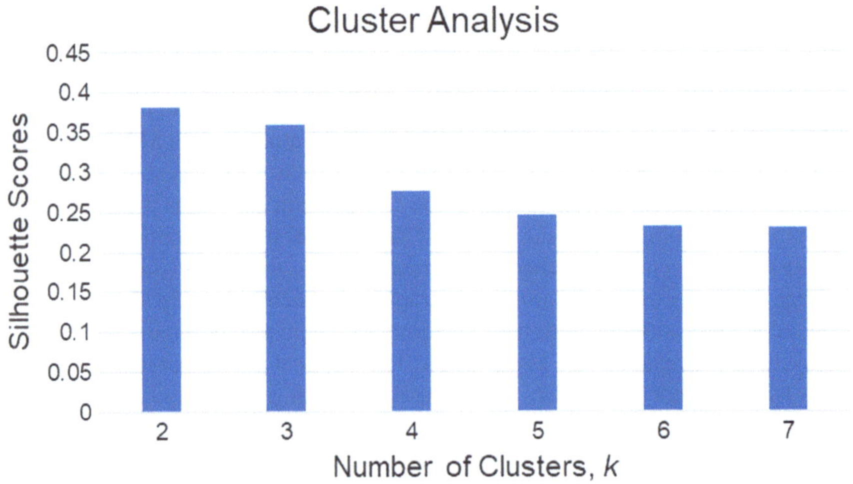

Fig. 5.1 Cluster analysis

5.4 Results and Discussion

Figure 5.2 depicts the box plot classes of the shooting score associated psycho-fitness variables measured in the present study. It could be observed from the plots that the HPA outperforms the LPA with regard to all the measured variables except for static balance which appears not to distinguish well the LPA and HPA. Therefore, from the

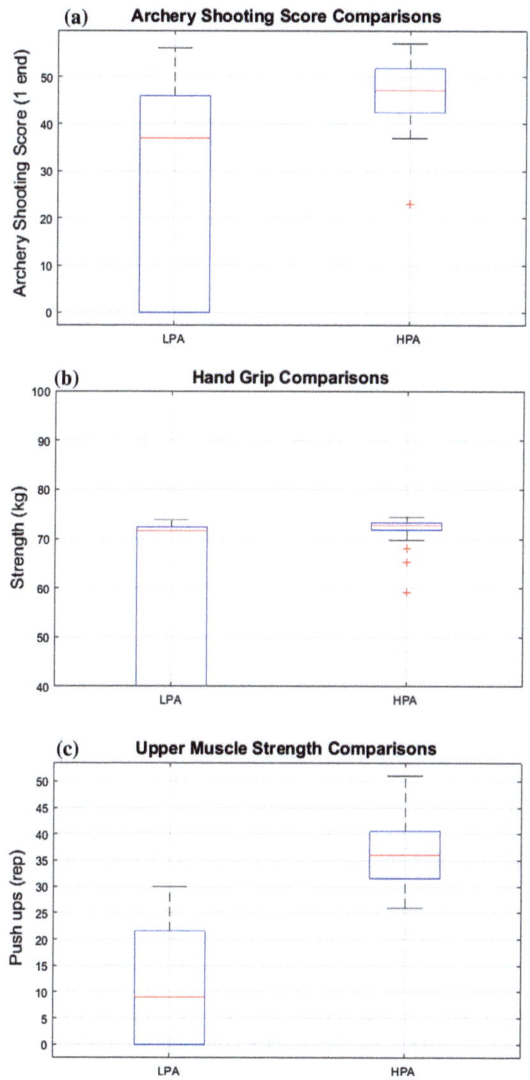

Fig. 5.2 Psycho-fitness box plot classes. **a** Archery shooting score, **b** hand grip, **c** upper muscle strength, **d** core muscle strength, **e** static balance, **f** vertical jump and **g** athletic coping Skills

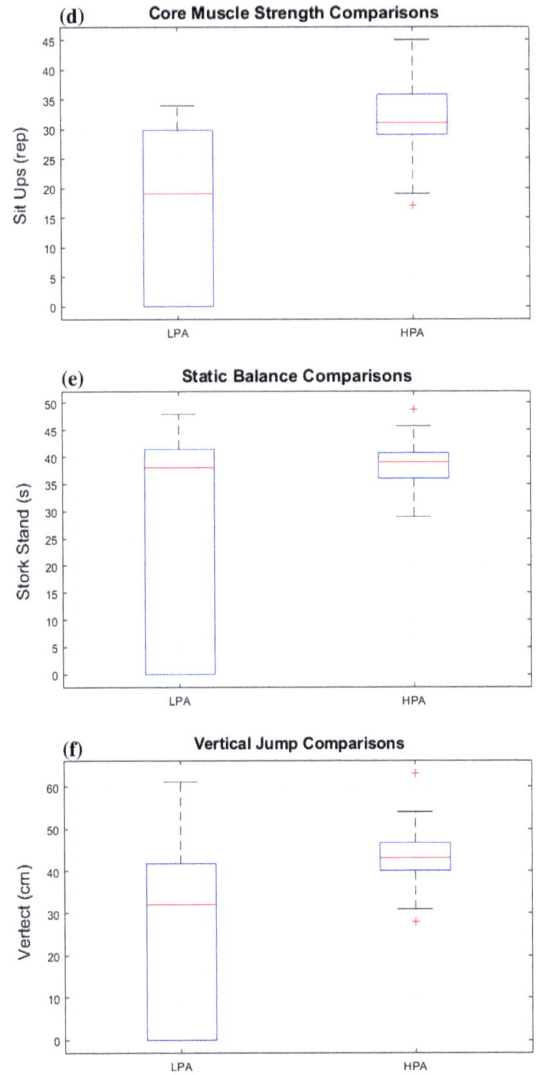

Fig. 5.2 (continued)

analysis it is evident that the selected features are non-trivial towards the performance of the sport of archery.

It is clear from Table 5.1 that the cosine k-NN variation classifier provided the highest classification accuracy of 82%. Moreover, it also performs reasonably well with respect to other evaluation metrics. Although both the fine and the weighted k-NN variation contributed to an equal classification accuracy of 80%, nonetheless, it is worth noting that other performance metrics differ. On the other hand, the medium

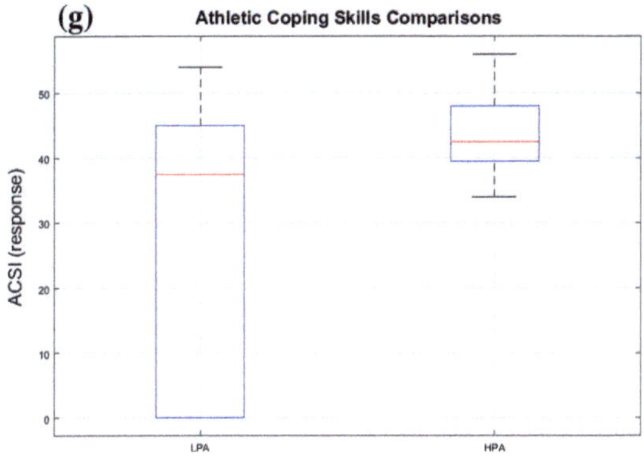

Fig. 5.2 (continued)

Table 5.1 Model evaluation

Algorithms	ACC (%)	SENS (%)	SPEC (%)	PREC (%)	ERR (%)	MCC
Fine *k*-NN	80.00	80.65	78.95	86.21	20.00	0.5861
Medium *k*-NN	70.00	80.65	52.63	73.53	30.00	0.3463
Coarse *k*-NN	62.00	100.00	0.00	62.00	38.00	N/A
Cosine *k*-NN	82.00	90.32	68.42	82.35	18.00	0.6113
Cubic *k*-NN	70.00	87.10	42.11	71.05	30.00	0.3319
Weighted *k*-NN	80.00	87.10	68.42	81.82	20.00	0.5689

and cubic *k*-NN variation yields a classification accuracy of 70%, respectively. The coarse *k*-NN model is deemed to be the worst performing classifier observed from the analysis. This does not come as a surprise, as all the LPAs were misclassified as HPA as shown in the confusion matrix in Fig. 5.3 that in turn resulted in an uncorrelated MCC.

The results from the present study have clearly indicated that the need for psycho-fitness parameters in the sport of archery cannot be overemphasised. It has been shown from the study that high-potential archers are physically fitter as well as more psychologically equipped in comparison with the low-potential archers as shown in Fig. 5.2. The findings from the study have also illustrated that the psycho-fitness parameters examined in the study could predict the performance classes of the archers virtually HPA and LPA.

A number of literature have reported the necessity for the acquisition of physical fitness variables such as upper and core muscle strength, leg power as well as hand gripping ability [1, 4]. These sets of physical fitness variables could provide the archer with the ability to cater for the rigorous tasks of carrying the bow and arrow, shooting, retrieving the arrows as well as standing firm for the long hour involves

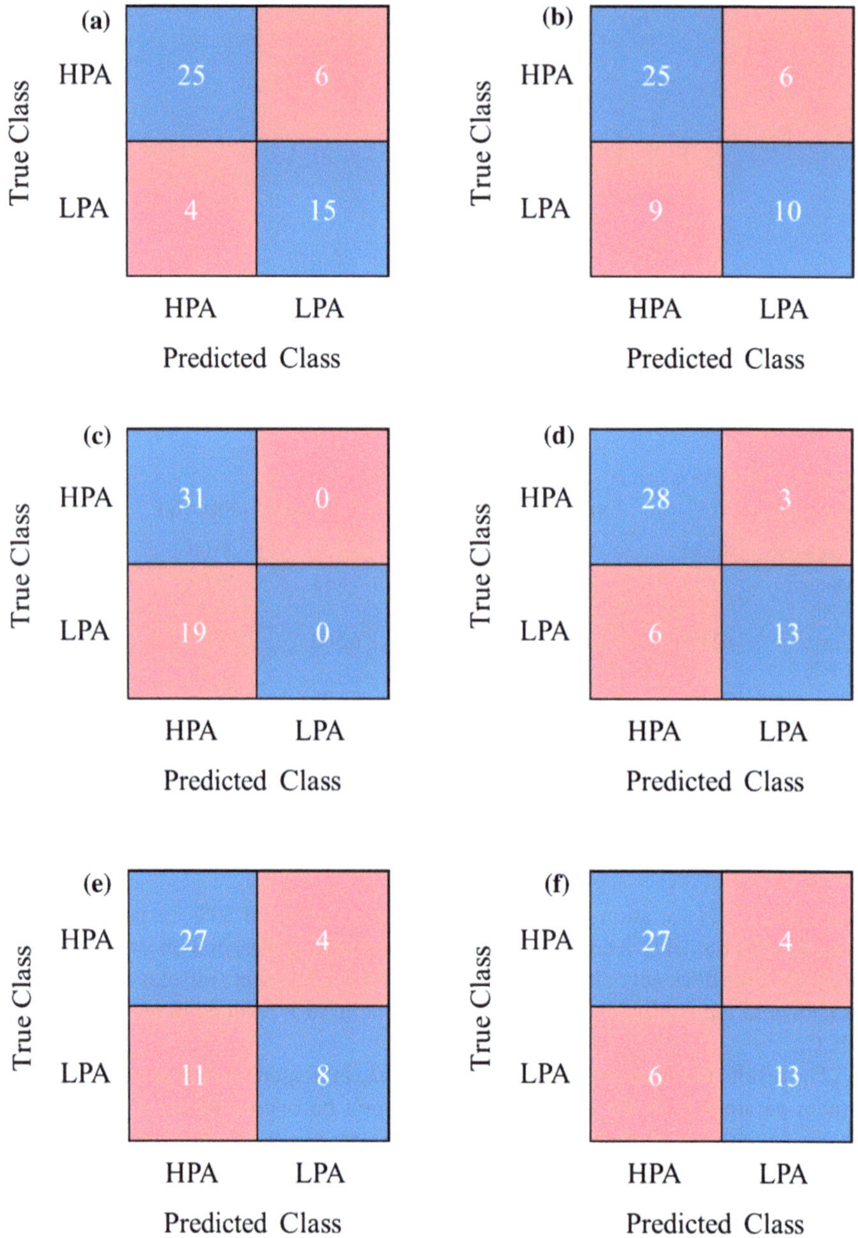

Fig. 5.3 Confusion matrix. **a** Fine k-NN model, **b** medium k-NN model, **c** coarse k-NN model, **d** cosine k-NN model, **e** cubic k-NN model and **f** weighted k-NN model

in performing the sport. It has also been reported that the capacity to maintain static balance in the course of archery is reliant on the level of strength and endurance of the limbs [5]. Since the sport typically constitutes a constant shooting and continues movement in the range, as such the need for the endurance of the muscles as well as the strength of both the upper and lower limb is of paramount importance for the attainment of success.

In another perspective, psychological coping skills strategies have been reported to be a significant determinant in ensuring success in the sporting domain [6]. How an athlete deals with stress during and after a sporting event could influence the delivery of performance and consequently affect the outcome of the match event. In the sport of archery, stress, anxiety as well as pressure have constantly been surrounding the sport, as such an archer that has the ability to deal with stressors is found to be excellent in their performance and thus is linked with higher achievement in the sport [7, 8]. It is, therefore, crucial for archers to have an optimum degree of psycho-fitness parameters in order to meet the psychological as well as the physical demands that are involved in the sport.

5.5 Summary

The present investigation has evaluated the contributions to the psycho-fitness parameters towards the successful performance of archery sport. The study has also demonstrated that potential archers could be identified based on the utilisation of the psycho-fitness parameters. It has been shown that high-potential archers are inherent with a comparatively higher degree of psycho-fitness parameters as compared to the low-potential archers. Furthermore, it has been demonstrated that the cosine k-NN variation provided a reasonable classification accuracy based on the chosen variables. It is, therefore, concluded that for the attainment of a higher score in the sport, the acquisition of the aforesaid parameters is non-trivial.

References

1. P.K. Suppiah, R.M. Musa, T. Wong, K. Kiet, M.R. Abdullah, A. Bisyri, H.M. Maliki, N.A. Kosni, M. Haque, Sensitivity prediction analysis of the contribution of physical fitness variables on Terengganu Malaysian youth archers' shooting scores. Int. J. Pharm. Sci. Rev. Res. **43**, 133–139 (2017)
2. R.M. Musa, M.R. Abdullah, A.B.H.M. Maliki, N.A. Kosni, M. Haque, The application of principal components analysis to recognize essential physical fitness components among youth development archers of Terengganu, Malaysia. Indian J. Sci. Technol. **9** (2016)
3. R.M. Musa, M.R. Abdullah, A.B.H.M. Maliki, N.A. Kosni, S.M. Mat-Rasid, A. Adnan, H. Juahir, Supervised pattern recognition of archers' relative psychological coping skills as a component for a better archery performance. J. Fundam. Appl. Sci. **10**, 467–484 (2018)

4. Z. Taha, R.M. Musa, A.P.P. Abdul Majeed, M.M. Alim, M.R. Abdullah, The identification of high potential archers based on fitness and motor ability variables: a support vector machine approach. Hum. Mov. Sci. (2018)
5. Z. Taha, R.M. Musa, M.R. Abdullah, M.A.M. Razman, C.M. Lee, F.A. Adnan, M.A. Abdullah, M. Haque, The application of inertial measurement units and wearable sensors to measure selected physiological indicators in archery. Asian J. Pharm. Res. Heal. Care. **9**, 85–92 (2017)
6. Devonport, T., Understanding stress and coping among competitive athletes in sport. Sport Exerc. Psychol. **127** (2015)
7. J.N. Gilbert, W. Gilbert, C. Morawski, Coaching strategies for helping adolescent athletes cope with stress. J. Phys. Educ. Recreat. Danc. **78**, 13–24 (2007)
8. H.A. Aysan, The effects of archery as a sports branch on coping with stress. Stud. Ethno-Medicine. **10**, 39–43 (2016)

Chapter 6
Concluding Remarks

The current brief evaluates the interactions of the various performance variables that effects performance in the sport of archery. Performance variables discussed in this brief included physical fitness constituting hand grip, static balance, leg power, upper and core muscle strength, physiological indicators involving nutrition, heart rate, blood pressure and psychological elements comprising of concentration, coping with adversity, coachability, confidence and achievement motivation, goal settings and mental preparation, peaking under pressure and freedom from worry. The physical characteristics or anthropometry of the athletes is also considered. It is worth noting that all the assessed variables have been shown to have an influence on the archery performance. The authors have utilised several machine learning algorithms for the purpose of identifying potential athletes within the sport based on the evaluated performance variables.

It has been established from the results that high- and low-performance archers could be discriminated against based on their performances on the selected evaluated performance variables. The identification of the potential archers was made possible through the application of various machine learning algorithms. Thus, it is worth to mention that the utilisation of such machine learning algorithms is non-trivial in the provision of an evident-based technique for the objective evaluation as well as the reasonable classification of the performance classes of the archers examined in the study. The technique proposes in the present brief serves useful to coaches, sports managers, sports scientist and the other stakeholders' concern for the identification of potential archers which could consequently reduce the energy, cost as well as the hassle involved in the conventional technique. Finally, it is important to highlight that the current findings could not be generalised to another level of expertise in the sport. The participants under the present investigation are senior youth archers as such the parameters that could be essential amongst the senior youth archer might differ from other level of expertise. However, the methods applied could be recommended for extension to other specific sport as well as various levels of sporting expertise.

R. Muazu Musa et al., *Machine Learning in Sports*, SpringerBriefs in Applied Sciences and Technology, https://doi.org/10.1007/978-981-13-2592-2_6

MIX
Papier aus verantwortungsvollen Quellen
Paper from responsible sources
FSC® C105338

If you have any concerns about our products,
you can contact us on
ProductSafety@springernature.com

In case Publisher is established outside the EU,
the EU authorized representative is:
Springer Nature Customer Service Center GmbH
Europaplatz 3, 69115 Heidelberg, Germany

Printed by Libri Plureos GmbH
in Hamburg, Germany